1.00

SCIENCE, SCRIPTURE
and the
YOUNG EARTH

**An Answer to Current Arguments against
the Biblical Doctrine of Recent Creation**

Henry M. Morris

Institute for Creation Research
El Cajon, California

SCIENCE, SCRIPTURE AND THE YOUNG EARTH

Copyright © 1983

Henry M. Morris

Library of Congress Catalog Card Number 83-81187

ISBN 0-932766-0604

Cataloging in Publication Data

Morris, Henry Madison, 1918 -

Science, scripture and the young earth: an answer to current arguments against Biblical doctrine of recent creation/by Henry M. Morris.

1. Bible and geology.   2. Creation.
3. Bible and science.

215   83-8117

ISBN 0-932766-06-4

Printed in the United States of America

SCIENCE, SCRIPTURE AND THE YOUNG EARTH

Introduction

Dr. Davis A. Young, a geology professor at Calvin College, has for some time been one of the most vocal opponents of the modern revival of scientific Biblical creationism, even though he is an avowedly evangelical Christian.

His main efforts in this direction have been in producing two books, which already have had a deleterious influence among many evangelicals. The first of these was *Creation and the Flood* (Grand Rapids, Baker Book House, 1977, 217 pp.). This book attempted especially to expound Biblical reasons for embracing the day/age theory. The book made it obvious, however, that it was not because of his Biblical exegesis that he had capitulated to uniformitarianism, but the other way around. His decision to go along with the geological establishment had compelled him to find some means, devious though it must be, to justify his position Biblically, following the same route that many theological accommodationists before him had traversed. This writer prepared a brief critical review of Young's first book,[1] but the Biblical expositions in the book were so obviously *ad hoc*

---

[1]See *The King of Creation*, by Henry M. Morris (San Diego: Creation Life Publishers, 1980), pp. 86-95.

that its influence was limited generally to those who already held a rather weak view of Scripture. Its scientific discussions were also superficial, and the book as a whole seemed hastily thrown together (with neither index nor bibliography, for example).

Young's second book, on the other hand, exhibits a considerably higher degree of scholarship, at least in several of its chapters, and will probably have a wider influence than the first book. This new book is entitled *Christianity and the Age of the Earth* (Grand Rapids, Zondervan Publishing House, 1982, 188 pp.). It deals only minimally with Biblical exposition, referring the reader back to his first book for this. It especially attacks the doctrines of a young earth and "flood geology," using scientific arguments almost exclusively. It is much better documented than Young's earlier book, and does contain good indexes of names and subjects.

Since this book undoubtedly will make a greater impact than the first book, it requires a more extensive review and critique than the first one did. On the other hand, it is impracticable to write a critique as long as the book itself (as would be required to deal adequately with the arguments advanced by Young), so only the more important aspects will be discussed.

Dr. Young first reviews church history in relation to the age question, throughout the first five chapters, occupying the first 40% of the book. The section on scientific considerations occupies four chapters and 40%, the last section, on philosophic considerations, has two chapters and 20%.

Beliefs of the Early Christians

The historical section starts with ancient Greek views on fossils and the age of the earth, continues through those of the early church, and eventually discusses modern Christian interpretations, including what he calls the "reactionary developments" of the modern creation movement. He does render one distinct service in this connection, showing that all the early Christian expositors believed in a young earth. This view was held, of course, on the basis of Biblical exegesis alone, since the prevailing view among the Greeks, the Egyptians, the Babylonians and other advanced nations of the world at the time, was that the world was very old, probably eternal. Some modern Christians have tried to find the day/age theory or the gap theory in the writings of Augustine or other early Christians, but Young shows that all of these men believed in a young earth.

He also shows that the church writings of the medieval and reformation periods likewise adhered to a young earth, literal creation days and a worldwide flood. This was especially true of great expositors such as Martin Luther and John Calvin. All of this effectively demonstrates that the writings of Scripture, taken at face value, clearly teach recent creation and a global flood.

Therefore, the evidence for an old earth must come exclusively from extra-Biblical sources. Even Young admitted, in *Creation and the Flood* that the literal-day interpretation of Genesis not only is a "legitimate" interpretation (p. 44) of the Biblical text, but also is "the obvious view" (p. 48). He thus has had to devise a "non-obvious" interpretation of Genesis to accommodate the geological interpretations which, for him, constitute the determinative criteria in his belief system. He (as do other accommodationists, past and present) argues that the Genesis record is sufficiently elastic so that we do not have to interpret it literally (that is, we do not have to believe that it means what it says!). We can allow "day" to mean "age," allow the days to overlap each other indiscriminately, let "evening and morning" mean "age-ending and age-beginning," let the great lights only "appear" to be set in the heavens, and understand God as still "resting" (in order to interpret the seventh day as an age) but also as still "creating and making" in order to justify extrapolating present processes into the creation by the principle of uniformitarianism.

In any case, Young makes it plain that his real reasons for holding the long-age view are geological, rather than Biblical. In so doing, he renders those of us who believe the Biblical record is inerrant, authoritative, and perspicuous a real service. The data of geology, in our view should be interpreted in light of Scripture, rather than distorting Scripture to accommodate current geological philosophy.

The Biblical Model of Historical Geology

In the second section of *Christianity and the Age of the Earth* is found the real essence of the book, namely, Dr. Young's scientific arguments against the creationists. He maintains first, in a chapter entitled "Stratigraphy, Sedimentation and the Flood," that the great thicknesses and varieties of sedimentary rocks require long ages for their deposition and lithification.

On the other hand, he agrees with creationists that catastrophism is a very prominent feature of the geologic column. Fossil graveyards, polystrate fossils, preservation of soft parts, and other such evidences of catastrophism cited in creationist writings are acknowledged by Young to have been produced by catastrophic burials. He maintains, however (creationists, of course, have always acknowledged this anyway) that uniformitarianism *includes* local catastrophes in its framework, so that these do not *necessarily* argue for a worldwide flood.

But the point is that *all* geologic systems and structures in the geological column testify of catastrophism! More and more geologists today are acknowledging this fact, even though they are not yet ready in most cases to become creationists. Dr. Derek Ager, Head of the Geology Department at Swansea University, England, and past president of the British Geological Association, has devoted an entire book to expounding this fact. After discussing a wide variety of geologic formations and features — all illustrating catastrophism — Dr. Ager concludes his book as follows:

> In other words, the history of any one part of the earth, like the life of a soldier, consists of long periods of boredom and short periods of terror.[2]

---

[2]Derek Ager, *The Nature of the Stratigraphical Record*, 2nd Ed. (New York, John Wiley Publishers, 1981),pp. 106, 107.

That is, in any single *local* geologic column, what we have is a record of several geologic catastrophes, and *no other record!* Ager, of course, does not believe these were all the *same* catastrophe, but different ones, scattered throughout geologic time.

I am coming more and more to the view that the evolution of life, like the evolution of continents and of the stratigraphical column in general, has been a very episodic affair, with short 'happenings' interrupting long ages of nothing in particular.[3]

But I maintain that a far more accurate picture of the stratigraphical record is of one long gap with only very occasional sedimentation.[4]

This is not merely a British view of things. In his Presidential Address (June 1982) to the American Society of Economic Paleontologists and Mineralogists, Professor Robert Dott of the University of Wisconsin, co-author of one of the nation's most widely used textbooks on general geology, said:

I hope I have convinced you that the sedimentary record is largely a record of episodic events rather than being uniformly continuous. My message is that episodicity is the rule, not the exception."[5]

Dott's reason for using the word "episodic," rather than "catastrophic" (which is what he means) is very interesting.

What do I mean by 'episodic sedimentation?' Episodic was chosen carefully over other possible terms. 'Catastrophic' has become popular recently because of its dramatic effect, but it should be purged from our vocabulary because it feeds the neo-catastrophist-creation cause."[6]

Not only should creationist/catastrophist teachings be banned from the schools and textbooks, but even creationist *words*! This is the noble scientific attitude to which Young and

[3]*Ibid.*, p. 99.

[4]*Ibid.*, p. 35.

[5]Robert H. Dott, "Episodic View Now Replacing Catastrophism," *Geotimes*, November 1982, p. 16.

[6]*Ibid.*

other evangelical accommodationists feel we Christians should all defer.

Nevertheless, the important point is that some leading geologists today are finally recognizing that the geological column is almost exclusively a record of *catastrophe,* not *uniformity*! They may still believe in "long periods of boredom," or "long ages of nothing in particular" or "long gaps," but when they do, they are simply exercising blind faith in long ages. *Science* is "knowledge," "observation," "experimentation," what we *see* — not what we don't see! And what we see in the geological record, everywhere and always, are evidences of catastrophe and rapid processes, not long ages of slow, continuous sedimentation.[7]

Now some uniformitarian die-hards, such as Davis Young, still maintain that some formations must have required long periods of time. He particularly mentions: (1) coral reefs; (2) evaporites; (3) lake deposits; (4) glacial deposits; and (5) desert deposits. These are admittedly the most difficult to interpret catastrophically, but it is at least *possible,* if not preferable, to interpret them this way.

Coral Reefs and Evaporites

As far as coral reefs are concerned, it should be realized that fossil "reefs" are probably not reefs at all. Creationist geologist Stuart Nevins (whom Young cites, but apparently without comprehension of his arguments), showed some time ago that one of the greatest of these, the El Capitan Permian "reef complex" in west Texas, was not a true reef, but largely an "allochthonous" (transported into place from elsewhere) deposit of fossil-bearing lithified lime mud.[8] Nevins gave an abundance of evidence in support of this conclusion, all of which Young ignores.

---

[7] For a thorough recent indictment of uniformitarianism in a secular journal, see J.H. Shea, "Uniformitarianism and Sedimentology," *Journal of Sedimentary Petrology* (Volume 52, 1982), pp. 701-702.

[8] Stuart E. Nevins, "Is the Capitan Limestone a Fossil Reef?" in *Speak to the Earth,* ed. by G.E. Howe (Philadelphia: Presbyterian and Reformed Co., 1975), pp. 16-59.

However, other leading geologists - not creationists - have come to similar conclusions about this and other fossil reefs. Speaking of the above-mentioned reef complex, Braithwaite says:

There is little doubt that this 'reef' had real topographic expression and that it controlled the distribution and character of sediments and organisms in the region. What does seem questionable is that this was an organic reef in which bioconstruction determined the feature. Dunham (1970) stated that the structure was in fact largely the product of inorganic binding, and that organisms, although present, did not provide a rigid framework."[9]

The author of this paper was Professor of Geology at Dundee University.

A true coral reef contains the binding framework of the coral organisms themselves. Fossil reefs, however, are "inorganically" bound, not "bioconstructed." That is, the evidence indicates that coral and other fossil organisms were simply transported into place by sedimentary processes, with the lime muds which constituted the matrix of transport later becoming lithified to form the present fossiliferous limestones.

Closer inspection of many of these ancient carbonate 'reefs' reveals that they are composed largely of carbonate mud with the larger skeletal particles 'floating' within the mud matrix. Conclusive evidence for a rigid organic framework does not exist in most of the ancient carbonate mounds. In this sense, they are remarkably different from modern coral reefs."[10]

As far as present living reefs are concerned, these have been shown to consist of a more or less superifical veneer of true coral growing on the surface of a non-coral base.

MacNeil (1954) put forward the idea that much of the

[9]C.J.R. Braithwaite, "Reefs: Just a Problem of Semantics?," *Bulletin, American Association of Petroleum Geologists* (Vol. 57, June 1973), p. 1105.

[10]H. Blatt, G. Middleton and R. Murray, *Origin of Sedimentary Rocks* (Englewood Cliffs, N.J., Prentice-Hall, 1972), p. 410.

morphology of modern reefs is explicable in terms of growth of corals on a pre-existing base, the shape of which was largely responsible for the form which the reef later took."[11]

In summary, present reefs commonly are bedded structures. . . Coral frames commonly represent only a small part of the volume."[12]

There is need for further research on the formation of reefs. but it is obvious that long ages are certainly not necessary for the construction of so-called coral reefs, either living reefs or fossil reefs.

Evaporites are even more clearly the result of rapid processes, in spite of their very misleading name. The typical uniformitarian explanation, repeated by Young, is that these great thicknesses of salt beds, gypsum beds and the like, have been formed by slow, and cyclically repeated, evaporation from inland lakes or relict seas. Such an explanation, however, is increasingly recognized today as impossible. The most extensive treatment of this subject is found in a monograph by the Russian geologist V.I. Sozansky.[13] In an English-language review of this book, the reviewer writes:

In Sozansky's book new data on the geology of world saliferous basins are compiled and summarized and result in the conclusion that, contrary to the classis conception of evaporitic origin, salt arises along faults as juvenile hot brines from the mantle."[14]

Sozansky comes to the conclusion that basins of salt accumulation were in active tectonic relation to depressions of block structure in which volcanic eruptions were common. Salt is not an evaporitic formation or a derivative from volcanic rock; it is a product of degasification of the earth's in-

---

[11]Braithwaite, op. cit., p. 1108.

[12]Ibid.

[13]V.I. Sozansky, Geology and Genesis of Salt Formations (Kiev, Izd. Naukova Kumka, 1973), 200 pp.

[14]V.B. Porfir'ev, Bulletin of the American Association of Petroleum Geologists (Volume 58, December 1974), p. 2543.

terior. The salt precipitated from juvenile hot water which emerged along deep faults into a basin as a result of change in thermodynamic conditions."[15]

Sozansky gives some of the reasons for rejecting the uniformitarian explanation of "evaporites," as follows:

The absence of remains of marine organisms in ancient salts indicates that the formation of the salt-bearing sections was not related to the evaporation of marine water in epicontinental seas. Other geologic data, such as the great thickness of salt deposits, the rapid rate of formation of salt-bearing sections, the presence of ore minerals in salts and in the caprocks of salt domes do not conform with the bar hypothesis."[16]

The analysis of recent geologic data, including data on the diapirs found in ocean deeps, permits the conclusion that these salts are of a juvenile origin — that they emerged from great depths along faults during tectonic movements. This process is often accompanied by the discharge of basin magmas."[17]

Needless to point out, this much more reasonable explanation of the origin of salt deposits is not only catastrophic in nature but completely consistent with the flood model of geology.

Lakes, Glaciers and Deserts

Next, Young insists that so-called lacustrine deposits require long periods of time, especially those containing "varves," or cyclic annual deposits, presumably formed seasonally along the beds of large lakes. He particularly stresses the famous Green River oil shale formation which covers a large area in Wyoming and adjacent states. This formation contains several million very thin layers which, according to uniformitarian thinking, were

---

[15]*Ibid.*, p. 2544.

[16]V.I. Sozansky, "Geological Notes. Origin of Salt Deposits in Deep-Water Basins of Atlantic Ocean," *Bulletin, American Association of Petroleum Geologists* (Volume 57, March 1973), p. 590.

[17]*Ibid.*

varves laid down at the rate of one each year, the whole formation thus requiring millions of years.

As a matter of fact, this formation (as well as other so-called lacustrine deposits in the geological column) has been the subject of considerable controversy even among orthodox geologists. For one thing, it is to a considerable extent a "fossil graveyard" and, as even Young admits, extensive beds of vertebrate fossils certainly imply catastrophic burials. Not only are extensive beds of fossilized fish found there, but even great deposits of fossil birds, as well as other animals.

> The discovery of abundant fossil catfish in oil shales in the Green River formation. . . .[18]

> Furthermore, fossil catfish are distributed in the Green River basin over an area of 16,000 km[2]. . . . The catfish range in length from 11 to 24 cm., with a mean of 18 cm. Preservation is excellent. In some specimens, even the skin and other soft parts, including the adipose fin, are well preserved. . . . ~~strongly suggests that the catfish could have been transported to their site of fossilization.~~[19]

> Because most bird bones are hollow or pneumatic as an adaption for flight, they are not well preserved in the fossil record. . . . During the early to mid-1970's enormous concentrations of Presbyornis have been discovered in the Green River Formation.[20]

Presbyornis is an extinct shorebird, equipped with a mosaic of the features of ducks and flamingoes. The fact that "abundant" fossil fish and "enormous concentrations" of fossil birds are found in the Green River formation surely ought to satisfy

---

[18]H. Paul Buchheim and Ronald C. Surdam, "Fossil Catfish and the Depositional Environment of the Green River Formation, Wyoming," Geology, Vol. 5, April 1977, p. 196.

[19]Ibid., p. 198.

[20]Alan Feduccia, "Presbyornis and the Evolution of Ducks and Flamingoes," American Scientist (Volume 66, May-June 1978), pp. 298-299.

anyone that this is not a varved lake-bed formation at all,[21] but a site of intense catastrophism and rapid burial.

Young also considers glacial deposits to be a problem for creationists. As a matter of fact, however, most creationists accept the so-called Pleistocene Epoch, or Ice Age, as a real period in earth history, following the great flood. The problem is not with the Pleistocene glacial deposits, which are accepted by both creationists and evolutionists, but with the much more equivocal evidences of earlier ice ages. Young lists a number of evidences of Permian Age glacial deposits, and others have found similar evidences in the Precambrian.

However, it should be stressed that these deposits are, indeed, much more equivocal than the types of deposits which are accepted as evidence of the Pleistocene Ice Age. Young mentions striated bedrock and conglomerates as glaciation indicators, but such phenomena can be produced by other causes than glaciers. Especially could this be true in the complex of violent phenomena unleashed at the time of the great flood. Conglomerates usually indicate intense flood transport, and grooves in rocks could well be formed by either boulder-laden flood waters, landslides or tectonic convulsions. Although it may not be possible as yet to deduce exactly how these phenomena were produced in the violent and varied phenomena of the Flood, the evidence of catastrophism is clear enough. For example, 55 billion cubic meters of coarse sedimentary rock in Australia, formerly interpreted as a "tilllite," originally deposited in an ancient glacial period, were more recently shown to have been formed by subaquaeous mud-flows.[22]

---

[21]Even in modern lakes, the so-called "varves" may well be formed by catastrophic turbid water underflows, with many being formed annually. See A. Lambert and K.J. Hsu, "Non-Annual Cycles of Varve-like Sedimentation in Walensee, Switzerland," *Sedimentology* (Vol. 26, 1979), pp. 453-461.

[22]J.F. Lindsay, "Carboniferous Subaquaeous Mass-movement in the Manning-Macleay Basin, Kempsey, New South Wales," *Journal of Sedimentary Petrology* (Volume 36, 1966), pp. 719-732.

Finally, Young cites desert formations as contradicting the flood model, though he says little about them. If real desert-formed features do exist in the geologic column, this could indeed be a problem for the Biblical model, since the antediluvian environment was said by God to be all "very good," and the future promised restoration of these good conditions to the earth includes desert reclamation (e.g., Isaiah 35).

However, Dr. Young apparently does not realize that the main examples he cites, the sandstones of the Colorado Plateau, are objects of considerable controversy among orthodox geologists, with many convinced they are water-laid, rather than wind-laid formations.[23] The sharp cross-bedding noted in some of these can be produced by violent water action.[24] Also, these sandstones contain interbedded mudstones and siltstones, with some of the best known fossil dinosaur graveyards in North America, and it is extremely unlikely that dinosaurs could have lived in a desert environment or that the fossil beds could have been formed in any way except by flooding.

In fact, the existence of wind-formed deposits in the geological column anywhere is highly suspect, except in Pleistocene and Recent deposits. Sand is formed by water erosion and transport. Even the great deserts in the modern world (e.g., Sahara, Mojave, etc.) were under water in the very recent past, geologically speaking. Modern sand dune areas are not being converted into sandstone — at least not in arid environments. Lithification of loose sediment into solid rock always requires the presence of a chemical cementing agent plus water.

There are innumerable evidences of catastrophism and rapid burial in the sedimentary strata, as geologists such as Ager and Dott have pointed out. Even those few deposits which Young

---

[23]W.E. Freeman and G.S. Visher, "Stratigraphic Analysis of the Navajo Sandstone," *Journal of Sedimentary Petrology* (Volume 45, 1975), pp. 651-668.

[24]L. Brand, "Field and Laboratory Studies on the Coconino Sandstone (Permian) Vertebrate Footprints and their Paleoecological Implications," *Paleogeography, Paleoclimatology, Paleoecology* (Volume 28, 1979), pp. 25-38.

thinks are diagnostic of slow formation in the traditional uniformitarian sense are also, as we have seen, quite amenable to catastrophic modeling. In fact the Biblical flood phenomena provide better explanations for them than uniformitarianism does.

Creationists recognize, of course, that there is much research yet needed (as well as re-interpretation of already available research) to provide a full explanation of the formations in every local geological column in terms of the Biblical model. However, a great deal has been accomplished. Young only refers to a small number of creationist writings (none dated after 1975) and obviously is unfamiliar with most of them.

Radiometric Dating Assumptions

The rest of the scientific section of Young's book deals with "absolute-age" measurements, both a defense of the few radiometric techniques used by evolutionists to give long ages and an attack on a few of the many methods used by creationists to indicate short ages. There are three main radiometric dating methods used to give long ages (and these are the only ones discussed by Young), namely, the uranium/thorium-lead method, the potassium-argon method, and the rubidium-strontium method. Each of these radioactive decay systems involves a very long "half-life," and thus will necessarily give great ages if they are measurable at all, and so they have been the only methods satisfactory to evolutionists, who require great ages to make their speculative evolutionary scenarios even remotely feasible.

Young acknowledges that all such methods require three fundamental assumptions: (1) constant half-life; (2) isolated system; (3) known initial boundary conditions. None of these assumptions can be proved correct, or even tested, since they involve conditions during vast aeons when no observers were present to test their validity. Nevertheless, Young argues (as do other uniformitarians) that they are reasonable if samples are selected carefully. Creationists maintain that they are not reasonable assumptions, and that systematic errors of gigantic magnitude are almost inevitable in them, no matter how carefully the samples are collected.

For example, it is probable that during the flood year and for some time afterwards, with the precipitation of the prediluvian water canopy and all sorts of other physical convulsions taking place, radioactive decay processes were sharply accelerated. Other global catastrophes which even uniformitarians have postulated (e.g., swarms of asteroids at the close of the Mesozoic Era) might have had a similar effect. Fred Jueneman (who is not a creationist) has at least called attention to this possibility.

The age of our globe is presently thought to be some 4.5 billion years, based on radio-decay rates of uranium and thorium. Such 'confirmation' may be shortlived, as nature is not to be discovered quite so easily. There has been in recent years the horrible realization that radio-decay rates are not as constant as previously thought, nor are they immune to environmental influences. And this could mean that the atomic clocks are reset during some global disaster, and events which brought the Mesozoic to a close may not be 65 billion years ago, but rather, within the age and memory of man."[25]

Another significant possibility is that radioactive decay rates, like all other decay processes, started out high and then gradually attenuated, slowing down to their present essentially constant values only in recent times. Radioactive decay processes, so far at least, can be described only as statistical processes, and there is certainly no known inherent reason why they could not have been higher - even immensely higher - in the past than they are now. Other known decay processes behave in this manner, so why not these? Thus, if for any reason, the decay rate has been decreasing exponentially with time, and even if it is now decreasing so slowly as to be almost impossible to measure, the calculated ages of radioactive minerals would be tremendously reduced.[26]

---

[25]Frederick B. Jueneman, "Secular Catastrophism," Industrial Research and Development, June 1982, p. 21.

[26]Theodore W. Rybka, "Consequences of Time Dependent Nuclear Decay Indices on Half-Lives," Impact Series No. 106, ICR Acts and Facts, April 1982, 4 pp.

However, the assumption of constant rate is not as critical as the assumption that the system involved is a closed and isolated system, so that any changes in internal composition would be caused only be radioactive decay of parent component to daughter component. The fact is, of course, that it is completely idealistic to think that any system is going to remain isolated for a billion years, especially in an earth that is - as modern geologists think - in a highly dynamic state, with moving and colliding continents, exuding magmas, subducting seafloors, tectonic upheavals, and downheavals, and frequent "local" catastophes. It is no wonder that most radiometric measurements and calculations turn out to be "discordant," at least until various "fudge factors" are employed to bring them into concordance with each other and with the geologic column.

For example, the "neutron capture" hypothesis of Melvin Cook suggests that the radiogenic lead isotope 207 (normally formed by decay of Uranium 235) could be formed from Lead 206 by capture of free neutrons in its vicinity, and Lead 208 (normally formed by decay of Thorium 232) from capture of free neutrons by Lead 207. By applying this type of reasoning, and supported by considerable data, Cook showed that practically all radiogenic lead in the earth's crust could have well been produced this way rather than by decay of uranium and thorium.[27]

Similar problems exist in maintaining potassium and rubidium minerals as closed systems. Potassium 40 is easily leached by groundwater and Argon 40 is a gas which can easily enter or leave a potassium mineral system. Young argues that argon is relatively inactive chemically and is therefore unlikely to enter such a mineral, but this does not seem true at least in the case of igneous rocks originally formed under water, which is the type of rock most commonly dated by this method.

In an attempt to establish criteria for obtaining reliable K-Ar dates, conventional K-Ar studies of several Deep Sea Drilling Project sites were undertaken. K-Ar dates of these rocks may be subject to inaccuracies as the result of seawater alteration.

---

[27]Melvin E. Cook, *Prehistory and Earth Models* (London, Max Parrish, 1966), pp. 23-62.

Inaccuracies also may result from the presence of excess radiogenic Ar-40 trapped in rapidly cooled rocks at the time of their formation."[28]

After discussing in detail the very significant errors in potassium dating that can result from these and other causes, Seidemann concludes:

Strong indication of the reliability of a conventional K-Ar date, such as its concordance with the dates of co-existing minerals, must exist before geologic significance can be attributed to it."[29]

That is, one should rely on a potassium-argon date only if it happens to agree with a date which has already been determined by some other method.

Occasionally, in fact, dates obtained from two or more methods do happen to agree, and this is taken to prove that the respective systems involved were, indeed, closed systems after all. However, in view of the rarity of such concordances, it may well be that, when they do occur, it is merely a fortuitous coincidence, to be expected occasionally merely by the laws of chance. Most published dates are from discordant data. Even more significantly, most discordant data are never published at all. It is often simply assumed that discordant dates indicate open systems, which are therefore unsuitable for dating purposes and thus can be ignored. A geologist friend who prefers to remain anonymous has collected the follwing inadvertent admissions from the geological literature.

In general, dates in the 'correct ball park' are assumed to be correct and are published, but those in disagreement with other data are seldom published nor are discrepancies fully explained."[30]

---

[28]David E. Seidemann, "Effect of Submarine Alteration on K-Ar Dating of Deep-Sea Igneous Rocks," *Bulletin of the Geological Society of America* (Volume 88, November 1977), p. 1660.

[29]*Ibid.*, p. 1666.

[30]R.L. Mauger, "K-Ar Ages of Biotites from Tuffs in Eocene Rocks of the Green River, Washakie and Uinta Basins," *Contributions to Geology, Wyoming University*, Vol. 15(1), 1977, p. 37.

In conventional interpretation of K-Ar age data, it is common to discard ages which are substantially too high or too low compared with the rest of the group or with other available data such as the geological time scale. The discrepancies between the rejected and the accepted are arbitrarily attributed to excess or loss or argon."[31]

Thus, if one believes that the derived ages in particular instances are in gross disagreement with established facts of field geology, he must conjure up geological processes that could cause anomalous or altered argon contents of the minerals."[32]

The Mississippian age for sample NS-45 cannot be correct because it is grossly inconsistent with the stratigraphic position of the lavas."[33]

Much still remains to be learned of the interpretation of isotopic ages and the realization that the isotopic age is not necessarily the geologic age of a rock has led to an over-skeptical attitude by some field geologists."[34]

This is an inherent uncertainty in dating young volcanic rocks: anomalies may be detected only by stratigraphic consistency tests, independent dating techniques, and comparison with the known time scale of geomagnetic reversals during the last 5 m.y. (Cox 1969)."[35]

[31]A. Hayatsu, "K-Ar Isochron Age of the North Mountain Basalt, Nova Scotia," Canadian Journal of Earth Sciences, Vol. 16, 1979, p. 974.

[32]J.P. Evernden and J.R. Richards, "Potassium - Argon Ages in Eastern Australia," Journal of the Geological Society of Australia, Vol. 9, No. 1, 1962, p. 3.

[33]C.M. Carmichael and H.C. Palmer, "Paleomagnetism of the Late Triassic North Mountain Basalt of Nova Scotia," Journal of Geophysical Research, Vol. 73, 1968, p. 2813.

[34]P.E. Brown and J.A. Miller, "Interpretation of Isotopic Ages in Orogenic Belts," Geological Society of London Special, No. 3, 1969, p. 137.

[35]R.L. Armstrong, "Late Cenozoic McMurdo Group and Dry Valley Glacial History, Victoria Land, Antarctica," New Zealand Journal of Geology and Geophysics, Vol. 21, 1978, p. 692.

The internal consistency demonstrated above is not a sufficient test of the accuracy of the age determinations; they must also be consistent within any age constraints placed on intrusion by fossils in the country rocks."[36]

The most reasonable age can be selected only after careful consideration of independent geochronologic data as well as field, stratigraphic and paleontologic evidence, and the petrographic and paragenetic relations. . . . In an effort to evaluate a discordant age sequence, therefore, the data are adjusted in one of several ways. . . until the lead-uranium and lead-lead ages are in agreement."[37]

The Initial Conditions

But still more important than the closed-system assumption, with all its attendant baggage of discordancies, is the assumption of the initial conditions. In order to make a meaningful age calculation, one has to be able to determine the amount of radiogenic daughter product (lead, argon or strontium, as the case might be) which has actually been derived by radio decay from the amount of parental component (uranium, potassium or rubidium) contained within the system. It must be assumed that the *initial* amount of such daughter product was either zero or some known amount which can be subtracted.

Now *this* assumption begs the whole question of creation. If God really created the world, as Young says he believes, why not be willing at least to consider the possibility that God created it all perfect at the beginning?

With respect to the basic elements and their own components (protons, neutrons, electrons), it would be most reasonable for these all to be in equilibrium with each other and proportioned in those amounts throughout the world which would be most beneficial to all creatures in God's economy,

[36]I.S. Williams, W. Compston, B.W. Chapell & T. Shirahase, "Rubidium-Strontium Age Determinations on Micas," *Journal of the Geological Society of Australia*, Vol. 22, No. 4, 1975, p. 502.

[37]L.R. Stieff, T.W. Stern & R.N. Eichler, "Algebraic and Graphic Methods for Evaluating Discordant Lead Isotope Ages," *U.S.G.S. Professional Paper 414-E*, 1963, E-1.

established so by His creative forethought and gracious omnipotence. It would be expected in such an economy that families of elements which later might be associated in a "decay chain" would likewise be associated together at creation. Wherever uranium was located throughout the earth's crust or mantle, there would also be some associated lead, for example, as well as all the intermediate products in the uranium-lead chain. However, many lead deposits would be placed without these associated elements, because there would be much use for lead in the future world and only limited use for uranium, thorium, radium, etc.

Now such a concept should not be a problem to Dr. Young, who says he believes in a Creator God. If God did not create a functioning, harmonious good world, as the Bible says He did, then would Dr. Young say that He created a chaotic, struggling, groaning world? Those of us who believe in an omniscient, omnipotent, loving, personal God do have a problem with *this* concept, and we find it difficult to understand why evangelical uniformitarians do not!

Consequently, to blandly assume that there was no Lead 206 with the Uranium 238 in the originally created uranium minerals, and no Argon 40 with the Potassium 40, is to deny even the possibility that these minerals were created in the beginning in the most probable proportions. And this simply begs the whole question of creation.

However, entirely apart from the question of *created* radiogenic "daughter" products is the very real possibility (in fact, the near certainty) that significant amounts of daughter products were, indeed, already incorporated with the "parents" at the time the minerals were emplaced in the rocks where they are now found. All of these radioactive minerals seem to have been introduced into the earth's crust, if not by creation, at least by magmatic flows up from the mantle below the crust, where they had already acquired (if not by creation, then by some process of nucleogenesis when the elements were formed in the primordial fireball) their "daughter" elements. They are now found primarily in igneous rocks - at least it is only igneous rocks that are believed suitable for dating by radiometry.

This situation is what is found whenever the initial condition assumption can be checked on rocks of *known* age -that is, on igneous rocks (e.g., volcanic lava rocks) that have been formed in very recent times. Young acknowledges that Argon 40 is frequently incorporated with Potassium 40 in volcanic extrusions, since this phenomenon has frequently been documented (note, e.g., the Seidemann article cited above). Even more significant is the fact that any uranium-lead date obtained on such a newly formed igneous rock always seems to be immensely older than the true age.[38] The evidence is clear that the parent uranium isotope had been associated with the daughter lead isotope in the mantle, long before the rock itself was formed, so that the rock may have an "apparent age" of, say, a billion years, when its true age is zero years. Now, since other igneous rocks have almost certainly been formed in a similar fashion, by the flow of a magma up from the mantle, and since radiometric dating is done only on igneous rocks, it seems that all such radiometric dates would be subject to this same basic error. All the multitudes of radiometrically determined ages published in the geological literature are thus fallacious ages, not representing the rock ages at all, but rather having to do with whatever process of nucleogenesis formed the chemical elements in the first place, and then with whatever process of planetary genesis formed the earth and its mantle in the second place.

Isochron Diagrams and Mixing

Now, although few geochronologists even consider the possibility that radiogenic "daughter" isotopes may actually have been *created* in association with their "parents," it is quite common now to recognize that they may have been incorporated somehow at the time of rock crystallization. Various techniques have been devised to try to estimate this initial condition, so that the amount of daughter product in the mineral can be corrected, leaving only the amount formed by radio-decay as an index to the age.

---

[38]Sidney P. Clementson, "A Critical Examination of Radioactive Dating of Rocks," *Creation Research Society Quarterly*, Vol. 7, December 1970, pp. 137-141.

The problem is that all such techniques involve still more assumptions which are non-provable, and none of them really deal with the key problem - namely, that the daughter product has most likely been with its parent ever since the time of their creation (or "nucleogenesis," if a more naturalistic term is preferred). The most popular of these techniques is the use of "isochron diagrams." These have been applied to all the main radiometric methods, but have been of principle use in rubidium-strontium dating, and this is the method Young seems to promote most vigorously as providing sure proof of an ancient earth.

The isochron diagram is based on a "whole-rock" sample, rather than on rubidium-strontium ratios in individual minerals. Each of the latter is assumed to be contaminated with a constant proportion of initial strontium, which proportion (in relation to the total strontium content in each respective mineral) supposedly permeated the entire magma as it was being emplaced and crystallized. For each mineral sampled in the rock mass, the total amount of Rubidium 87 and Strontium 87 is measured, and all are then plotted on a dimensionless scale, with

$$\frac{\text{Strontium 87}}{\text{Strontium 86}} \quad \text{plotted versus} \quad \frac{\text{Rubidium 87}}{\text{Strontium 86}}$$

This is the "isochron" diagram as usually defined. Their reference term, Strontium 86, is a non-radiogenic isotope of strontium. At the time of crystallization the initial ratio of $^{87}Sr/^{86}Sr$ is assumed to be constant everywhere. As minerals form, some will have larger amounts of ^{87}Sr, some lesser amounts, but all are assumed to have the same proportion of $^{87}Sr/^{86}Sr$. Similarly, some will have larger amounts of Rubidium 87, others lesser amounts. The latter will begin to decay into strontium, with the larger rubidium bodies naturally generating larger amounts of strontium.

If all goes well, the slope of the isochron so plotted should indicate the age since crystallization, and the intersection of the line with the vertical axis should give the initial value of $^{87}Sr/^{86}Sr$. The problem is, however, that once again all does not usually go well, because the assumptions are bad.

In the first place, the assumption of uniform initial strontium content throughout the entire rock mass is quite unjustified. Most such "whole-rocks" actually are mixtures of two or more magmas, so that the initial ratios of Strontium 87 to Strontium 86 will vary from mineral to mineral. Consequently an isochron plotted at zero age could well have a significant slope, merely indicating the fact that different isotope ratios existed throughout the rock and having nothing whatever to do with its age.

That this is not a mere quibble is proven by the fact that the petrological literature yields numerous examples of isochron ages which are rejected because they disagree with the geologic age or even with the radiometric ages obtained by other methods. For example:

> These results indicate that even total-rock systems may be open during metamorphism and may have their isotopic systems changed, making it impossible to determine their geologic age.[39]

Many isochrons thus turn out to be "pseudo-isochrons," as they are called when their slopes happen not to agree with the pre-determined ages for the rocks they represent. Sometimes they indicate ages far greater than the accepted age of the universe; sometimes they have even indicated negative ages! It is increasingly being recognized that mixing is a very common phenomenon, and this fact to all intents and purposes completely invalidates the method.

Dr. Russell Arndts, Professor of Chemistry at St. Cloud State University in Minnesota, and William Overn, nuclear and computer scientist, now co-director of the Bible Science Association, have written a number of incisive critiques of these techniques, demonstrating that mixing is at least a likely problem in all such isochrons. After citing many cases which had been recognized by the evolutionists as pseudo-isochrons, Arndts then found over 23 additional "ages" as published and accepted in the literature, as randomly sampled by him, which he was able to show conclusively were based on pseudo-isochrons

---

[39]G. Faure and J.L. Powell, *Strontium Isotope Geology* (New York, Springer-Verlag, 1972), p. 102.

produced by hitherto unrecognized mixing processes.[40] His well-justified conclusion is that all such isochrons are highly suspect from this cause alone. Randal Mandock has also produced an extensive critique of radiometric dating, based on his own surveys of the mixing phenomenon as well as other fallacies in both the theory and techniques of radiometric dating.[41]

However, even if *no* mixing were present, and a good isochron were obtained, the result is still quite meaningless as far as true age is concerned, for the reasons discussed earlier. The "daughters" are already incorporated with the "parents" in the elemental structure of the earth, before magmatic flow from the mantle takes place at all. That is, the "apparent age," as deduced from the isochron, may well be an "inherited age" from the mantle (really an "inherited *apparent* age," since the isotope ratios are not due to radio-decay but to nucleogenesis). This problem also applies when age determinations are attempted on sedimentary and metamorphic rocks.

One serious consequence of the mantle isochron model is that crystallization ages determined on basic igneous rocks by the Rb-Sr whole rock technique can be greater than the true age by many hundreds of millions of years. This problem of inherited age is more serious for younger rocks, and there are well-documented instances of conflicts between stratigraphic age and Rb-Sr age in the literature.[42]

The major source of error in Rb/Sr age determinations on whole rock samples of shale is the presence of inherited

---

[40]Russell Arndts, et al., "Radiogenic Isotopes, Straight Lines and the Mixing Model," Paper presented at the Atlantic Creation Convention, August 1981, 16 pp.

[41]Randal L.N. Mandock, *Scale Time Versus Geologic Time in Radioisotope Age Determination*, Master of Science Thesis, (San Diego: Institute for Creation Research Graduate School), August 1982, 160 pp.

[42]C. Brooks, D.E. James & S.R. Hart, "Ancient Lithosphere: Its Role in Young Continental Volcanism," *Science*, Vol. 193, September 17, 1976, p. 1093.

radiogenic strontium in the detrital minerals.[43]

These data suggest that great caution must be used when applying the Rb-Sr whole-rock technique to metamorphic rocks. This and other studies show that metasedimentary rocks can yield linear arrays of points on isochron diagrams, but that the age can have a variety of geological meanings and need not define the date of deposition or metamorphism of the sediment.[44]

This discussion has been fairly extensive because not only Young but also many other evangelicals (not to mention out-and-out evolutionists) have been much too impressed and intimidated by the claims of the radiometric daters.

Decay of the Magnetic Field

Creationists, of course, not only have shown that radiometric dating does not really prove an old earth, but they have also developed many scientific evidences that point positively to a young earth.[45] One of the most important of these is the decay of the earth's di-polar magnetic field, an argument developed and highly refined by Dr. Thomas G. Barnes, Professor Emeritus of Physics at the University of Texas (El Paso) and now Dean of the Graduate School at the Institute for Creation Research.[46]

This is a worldwide process (not a local process, as in a uranium mineral), accurately measured for over 145 years (not for

[43]P.R. Whitney & P.M. Hurley, "The Problem of Inherited Radiogenic Strontium in Sedimentary Age Determination," *Geochimica et Cosmochimica Acta*, Vol. 28, April 1964.

[44]M. Spanglet, H.R. Brueckner & R.G. Senechal, "Old Rb-Sr Whole Rock Apparent Ages From Lower Cambrian Psammites and Metapsammites, Southeastern New York," *Bulletin of the Geological Society of America*, Vol. 89, May 1978, p. 789.

[45]For example, the book *What is Creation Science?* by Henry M. Morris and Gary E. Parker (San Diego: Creation-Life Publishers, 1982,) lists 68 such evidences (pp. 239-259).

[46]Thomas G. Barnes, *Origin and Destiny of the Earth's Magnetic Field,* 2nd Ed. (El Cajon, CA: Institute for Creation Research, 1983), 131 pp.

just a few years, as for radio-active decay processes) not subject to environmental changes, since it is generated deep in the earth's interior (not like radioactive minerals, which are highly unreliable becuase of the open systems in which they function), and probably not subject to changes in decay rates, since the factors that control it cannot be affected by any known outside conditions. If any process should be a *reliable* indicator of the earth's age, *this* should be — and it indicates an upper limit for the age of about 10,000 years!

No wonder, therefore, that Dr. Young must try diligently to belittle its testimony.[47] He, along with others who have argued against it, insist that the magnetic field is maintained by some sort of dynamo action in the earth's core, rather than by decaying electrical currents in the core, as Dr. Barnes maintains. This dynamo, in turn, is said to support the idea of an oscillating magnetic field, which reverses itself periodically. By Barnes' calculations, on the other hand, when the field strength decays to zero, that's the end of it. The currents which sustain the field will have died out and, by the principle of energy conservation, there is no energy source to restore it once it is gone.

Dr. Young does acknowledge that the dynamo theory is "fraught with difficulties," that there is no known cause to explain it and no real evidence that a dynamo even exists. The only real basis for the dynamo theory is that it avoids the young-earth implications of the electrical current theory, which Barnes has shown to be strongly supported by all known data and by sound physics, and which neither Dr. Young nor other critics have even attempted to refute.

Young also refers to a number of archaeological and geological measurements of "paleomagnetism," which he thinks indicate that the earth's magnetic field strength was not significantly greater in the past than it has been in the present. Such data are obtained from magnetic particles which were "frozen" in the rocks or in archaeological construction materials, ori-

---

[47]See Chapter 8 of *Christianity and the Age of the Earth*, pp. 117-124.

ented in the direction of the magnetic field at the time they were formed. From the "natural remanent magnetism" preserved thereby in these materials, geophysicists think they can determine both the direction and intensity of the magnetic field at the time they were laid down. On the basis of such data, the widely influential concept has been promoted that the magnetic field in the past has fluctuated both in direction and intensity. Barnes' 145 years of decaying field, they say, is therefore, only a part of the down-phase of the cycle.

However, there is one big difference. The data used by Barnes were truly for the earth's dipole magnet. To get them, worldwide measurements had to be made and integrated over a considerable period of time and over the entire globe. No local or regional measurement of magnetic moment could ever be used to do this alone, since there are innumerable local magnetic fields which can strongly influence and affect measurements in any given region. Thus, to think that the remanent magnetism in a suite of rocks or an archaeological site could be used to determine the earth's overall dipole magnetic strength or direction at that time in history is naive extrapolation carried to extremes.

The analysis developed by Dr. Barnes is based on sound physics, careful calculations and solid data. The dynamo theory, and the fluctuating and reversing magnetic field concept (except on a localized basis), are purely *ad hoc* notions, with no sound basis in either theory or measurement. The only real reason for rejecting the first and accepting the second is that the first supports recent special creation, the second tries to salvage a bankrupt evolutionary uniformitarianism.

In the enlarged 1983 edition of his monograph, Dr. Barnes has effectively rebutted the various criticisms that have been directed against his analysis and has still further strengthened his theory by additional proof. Critics are challenged to find any fallacy in either the data or analysis if they can. And until they do, Dr. Barnes has come as close as possible to a really scientific determination of the age of the earth. The almost inescapable conclusion is that the earth is only a few thousand years old, Dr. Young to the contrary notwithstanding.

Other Evidences for a Young Earth

Now, of course, the earth's decaying magnetic field is only one evidence of its youth. There are scores of other worldwide processes and systems that likewise point to a young earth. All of these, or course, must depend on the same kinds of assumptions that must be applied to radiometric age calculations — namely, constant rate, closed system and known initial conditions. In almost every case, these assumptions are at least as realistic as in the case of the radiometric calculations. Consequently, most creationist scientists are convinced - entirely apart from Biblical revelation - that the great weight of scientific evidence favors a recent creation.

Dr. Young selects three of these processes to criticize: the influx of meteorites to the earth; the influx of meteoritic nickel into the crust; and the erosion and deposition of sediment from the continents. These processes are no more significant than scores of others and it is not clear why Young chose these to rebut, especially when his arguments are *ad hoc* and tentative. The absence of meteorites in the geologic column he attributes to subduction into the mantle; the absence of sedimentary materials he attributes to multiple recycling. These are all secondary hypotheses, of course. The primary fact is that the meteorites, nickel and sediments that the regular uniformitarian assumptions would predict from present rates are all missing. It would be better evidence of long ages if they could be found!

Philosophy of the Young Earth

The final section of Davis Young's book is devoted to what he called "Philosophical and Apologetic Considerations Related to the Age of the Earth." This consists of two chapters, the first of which is a defense of the modern philosophy of uniformitarianism and is a modified reprint of an article he had written for the *American Scientific Affiliation Journal.*

This article reviews the modern creationist critique of geological uniformitarianism and also outlines what might be called the uniformitarian approach to catastrophism among modern geologists. However, Young complains (erroneously) that creationists have misunderstood uniformitarianism and have not realized that geologists *do* accept local and regional catastrophes as explanations for many geological systems.

Creationists have indeed been pleased to see the development of this neo-catastrophist movement among geologists, because it is at least coming closer to Biblical catastrophism than the Lyellian uniformitarianism which dominated geological thought up until the last 20 years or so. Possibly because Young is still "young" and also possibly because of his graduate studies at Penn State, which has been one of the few universities willing to consider catastrophism seriously (as long as it was not Biblical catastrophism!), he does not realize what a grip doctrinaire uniformitarianism had on most geological thinking until very recently. When *The Genesis Flood*[48] was written, for example, (this is apparently the creationist book Young has read most carefully, and which he criticizes most in his own book), it was necessary to devote two chapters totalling 123 pages just to demonstrate that traditional uniformitarianism could *not* account for the main features of the earth's geology.

There were, to be sure, a number of geological voices raised from time to time warning against a too-rigid uniformitarianism, but these were generally ignored by the great majority of geologists. Lyellian uniformitarianism had, indeed, gained essentially universal acceptance, and it *was* necessary for creationists to protest vigorously against what has now come to be called "substantive uniformitarianism" (uniformity of process rates), as distinct from "methodological uniformitarianism" (uniformity of natural laws). The exposition in *The Genesis Flood* strongly upheld the latter type of uniformitarianism.

These latter two terms were apparently first promoted by Stephen Jay Gould, in an important 1965 paper (four years after *The Genesis Flood* was published). It probably was the paper that provided the chief impetus (along with several papers of Harlan Bretz advocating a catastrophic interpretation of the spectacular "scabland" topography of the northwest) which prompted the current rejuvenation of catastrophism in geo-

---

[48]John C. Whitcomb and Henry M. Morris, *The Genesis Flood* (Philadelphia: Presbyterian and Reformed Publishing Co., 1961), 518 pp.

logical thought. Gould recommended the abandonment of substantive uniformitarianism.

> Substantive uniformitarianism as a descriptive theory has not withstood the test of new data and can no longer be maintained in any strict manner.[49]

Geologists have lately been climbing on this new bandwagon of naturalistic catastrophism in significant numbers. How much of this new awareness of catastrophism may have been due to the attacks on Lyellian uniformitarianism by creationists is an open question (geologists would never admit *this*, of course!), but in any case they are often now writing the same sorts of things creationists have been saying for years, though still vigorously resisting *Biblical* catastrophism.

Actually it is Young who is arguing redundantly, not the creationists. He makes quite a point of stressing that modern Biblical creationists are actually uniformitarians, since we argue in terms of uniform physical principles, even during the period of the great flood. That is, we creationists are methodological uniformitarians, adhering to the uniformity of natural law.

But, of course, that is exactly what we have been saying all along, and we said it long before Gould did! That is, we have stressed that a true science of historical geology should be based on established physical laws, not on speculative extrapolation of modern process rates into the imaginary pre-historic past. The scientific laws of mechanics and thermodynamics, of hydraulics and chemistry, of biology and genetics, together with a comprehensive understanding of the statistical rate variations possible in processes influenced by many variables, provide the only sound basis for an approach to interpreting the record in the rocks. And such an approach will not lead to the long ages of evolutionary geology!

Furthermore, the Biblical record of history is a real record of history! It does describe a worldwide cataclysmic geological upheaval a few thousand years ago, and there is no justification at all for professing Christians such as Young not to apply *this*

---

[49]Stephen Jay Gould, "Is Uniformitarianism Necessary?," *American Journal of Science*, Vol. 263, March 1965, p. 226.

fact also in their study of earth history.

Still further, God's Word specifically tells us that not even methodological uniformitarianism can be validly applied to the study of the creation period. Although God's natural laws and principles may now be invariant (except in specific miracles) they are *God's* laws. He made them! And he did so as a termination to His works of creating and making all things, from which He *rested* after the six days (Genesis 2:1-3). He is now "upholding all things" (Hebrews 1:3) but then He was creating and making all things. His creative and formative operations are no longer operational, and so were not under those present laws and principles which *are* operational. Here is the greatest fallacy of uniformitarian geology. Geologists insist on applying uniformitarianism to the study of the creation period, when God Himself has clearly told them (e.g., Hebrews 11:3) it cannot be done!

Evangelistic and Apologetic Implications

The final chapter of *Christianity and the Age of the Earth* is the most innocuous in terms of substance but the most offensive in terms of tone. It is primarily a moralizing preachment against creationists, as Dr. Young patronizingly scolds us unenlightened fundamentalists for presuming to speak on a subject in which only he and his colleagues are qualified. Note a few of his pronouncements:

Creationists need to learn how to receive criticism when they are told that they have spoken on a matter about which they know relatively little (p. 151).

I am *not* accusing creationists of lying or deliberate distortion. No doubt they have honorable intentions, but if they continue to espouse their theories when other Christian brethren have repeatedly called attention to the falsity of their theories, they must be challenged to stop (p. 152).

Creationism must be abandoned by Christians before harm is done. The persistent attempts of the creationist movement to get their points of view established in educational institutions can only bring harm to the Christian cause (p. 163).

May I plead with my brethren in Christ who are involved in the young-Earth movement to abandon the misleading writing they provide the Christian public. I urge them to study geology more thoroughly (p. 163).

We need to remind our pleading brother that there are today literally thousands of scientists who have become young-earth creationists, and that many of these - probably most of these - possess scientific credentials fully equal to those of Dr. Young and his associates. Furthermore, many or most of us (including the writer) started out as "old-earth evolutionists," not "young-earth creationists." It was because both the facts of science and the clear teaching of Scripture compelled us to do so that we have gone through this traumatic change - not because we were naive traditionalists who blindly believed what we were told.

Dr. Young also expresses concern lest creationism keep students and scientists from coming to Christ. The facts, however, run in exactly the opposite direction. The various compromise positions advocated by Young and other such writers (theistic evolution, progressive creation, gap theory, day/age theory, etc.) have been dominant in Christendom ever since Darwin.

But this Darwinian century, this century of Christian compromise, is also the same century that has been marked by the loss of our public schools and state universities to humanism, of society to materialism, fascism and communism, of Christianity to liberalism and socialism, along with a host of other ills that can be traced directly to departure from the Christian, creationist faith of our founding fathers.

This writer has been closely and directly associated with the university world for 45 years, and it has certainly been his experience that evangelism is far more fruitful and its results far more lasting when it is conducted in a solid framework of creationism and Biblical authority than one of expediency and minimal commitment. We at the Institute for Creation Research have received large numbers of verbal and written testimonials from students (as well as many from faculty and practicing professionals - even many geologists!) of conversions, soul-winning and spiritual growth as a direct result of our creationist literature and seminars, and even our debates. Young-earth, flood-geology, literal-day, Biblical creationism not only has solid roots, but also produces good fruits!